DO OR DIE

A SUPPLEMENTARY MANUAL ON INDIVIDUAL COMBAT

SHOWING ADVANCED SCIENCE IN
BAYONET, KNIFE, JIU–JITSU, SAVATE
AND BOXING
FOR THOSE WHOSE DUTIES MAY LEAD THEM
INTO A

"TIGHT SPOT"

BY

LIEUT.-COL. A. J. DREXEL BIDDLE, U.S.M.C.R.

INSTRUCTOR OF INDIVIDUAL COMBAT, UNITED STATES MARINE CORPS
AND FEDERAL BUREAU OF INVESTIGATION, UNITED STATES
DEPARTMENT OF JUSTICE

With an Introduction by

COLONEL C. J. MILLER, U.S.M.C.

COMMANDING THE FIFTH REGIMENT, UNITED STATES MARINES

E P B M

ECHO POINT BOOKS & MEDIA, LLC
BRATTLEBORO, VERMONT

Published in 2017 by Echo Point Books & Media
Brattleboro, Vermont
www.EchoPointBooks.com

Originally published in 1937

Do or Die
ISBN: 978-1-63561-500-5 (casebound)
978-1-62654-161-0 (paperback)

Cover design by Adrienne Núñez

INTRODUCTION

THERE is greater need of training the individual soldier than ever before. The tendency in modern warfare to utilize cover and to spread out more and more in the attack, in order to escape the devastating fires of the defender has forced the infantryman to rely primarily on his own personal skill, agility and courage to get forward and close with the enemy. The old linear and mass attacks have been replaced by small groups of skirmishers, working forward with calculated boldness and stealth, seeking out the defender's weak spots, in order to assault his combat groups. New methods of attack will require that the infantryman be a skirmisher, marksman, athlete and fighter par excellence.

The fighting powers of the individual soldier have increased in importance while mere weight of numbers has lost much of its value on the fireswept battlefield of today. Not only must the infantryman create his own opportunities, but he must be imbued more than ever with an aggressive spirit and a confidence in his own superiority, because he must depend on his own resourcefulness in battle.

Whether or not we believe the bayonet is still worth retaining as a weapon, bayonet fighting or its refinement, bayonet fencing, as illustrated in this Manual, remains a part and parcel of the individual's training

(3)

as heretofore. No other form of training instills greater confidence in the prowess of the soldier or creates that self-determination and overwhelming impulse to close with the enemy than is fostered by bayonet training. Herein lies its great value.

Self-preservation remains the great law of Nature. It is the heritage of every soldier to know how to protect himself under all situations. The assault will often lead to personal contact with the enemy, when the individual must know how to destroy his opponent and at the same time protect himself.

Individual combat, whether it be in the form of boxing, knife fighting, hand-to-hand encounter, or the Græco-Roman, catch-as-catch-can and jiu-jitsu wrestling, develops a suppleness of body, an agility afoot, a quickness of eye and a coördination of mind and body that adds immeasurably to the self-reliance and courage of the soldier in the close-up encounter with the enemy.

Lieutenant Colonel A. J. Drexel Biddle, U.S.M.C.R., with his extraordinary background of experience and study, has contributed in this text a most valuable and practical analysis of individual combat for developing the soldier's fighting and physical attributes. This Manual combines the art of self-defense and illustrates the methods of attack that will enhance the individual's natural powers of destruction.

C. J. MILLER,
Colonel, U. S. Marines.

IMPRIMATUR

THE writer has been an ardent student of the art of self-defense in all of its branches. During the past twenty years he has pursued an intensive study and training in the use of fencing blades. In the course of these studies he has sought every opportunity to secure instruction and personal training under the most eminent authorities and experts in the United States and abroad. It is with the knowledge and experience so gained that he now undertakes the preparation of this Manual.

Grateful recognition is accorded to all who have contributed to his knowledge or collaborated in the preparation of this work. Special mention must be made of some whose contributions are of outstanding value.

The writer has been trained by able sword and bayonet instructors of the British Army, including Sergeant J. H. Dawkins, the sword instructor of the King's Royal Horse Guards in London; he has received special instruction in Bayonet Combat at the military training school of Gondrecourt, in France, and in sword and dagger in several Portuguese, Spanish and French Colonies.

Lessons in broadsword were received from the Broadsword Champion, M. Thomas, at the Cercle Hoche in Paris where the writer also received instructions in general swordsmanship from the famous former Inter-

(5)

national Sword Champion, M. Surget, also an instructor at the Cercle. Afterwards the writer pursued his fencing studies under the celebrated fencing Master, Mr. J. Martinez Castello in New York.

JEAN-MARIE SURGET

Lauréat du Tournoi international d'épée (1902 et 1908)
1er prix de la Poule des Prévôts (1903)
1er prix de la Poule de Gala de l'Académie d'Epée (1903)
1er prix, Poule d'honneur du Tournoi international (1912)

Many years ago the writer began his instructions under the teachings of a former American Fencing Champion who, a few years since, won the bayonet fighting championship of the world. This latter named gentleman is Major William J. Herrmann, P.M.T.C., to whom the writer is thankful for the knowledge of some of the bayonet and knife movements prescribed in this treatise. Major Herrmann conducts the famous William J. Herrmann Physical Cultural Institute in Philadelphia where special attention is given by the Major and his fine staff to the instruction of teachers in bayonet, knife, and sword fencing.

In 1935 the Fifth Regiment of Marines commanded by Colonel Charles F. B. Price, U.S.M.C., was stationed at Marine Barracks, Quantico, Virginia, as part of the Fleet Marine Force, under command of Major General Charles H. Lyman.

The enthusiastic interest of the First and subsequent Regimental Commanders of the Fifth Regiment in Training for Individual Combat brought about an invitation to the writer to come to Quantico personally to serve as instructor to the Fifth Marines: it was his privilege to do this in August and September of the year named.

The gratifying results attained during that period of training created the desirability of developing this type of training further, and Colonel Price suggested to the writer the preparation of this Manual to be used as a Guide in future instruction.

The writer is particularly grateful to the Colonel for that suggestion and for the encouragement and coöperation since extended by valuable suggestions and by personally directing the preparation of illustrations.

As this Manual would not have come into being save for Colonel Price's timely suggestions and valuable assistance, and because the Colonel is an outstanding commander and an enthusiastic advocate of training in Individual Combat in the Marine Corps, it is a special pleasure to be allowed to dedicate this Manual to

COLONEL CHARLES F. B. PRICE
United States Marines.

Changes of command following the manœuvres with the Fleet in 1935 took Major General Lyman to command of the Marine Barracks, Quantico, brought Colonel (now Brigadier General) James J. Meade to command of the First Marine Brigade, Fleet Marine Force, took Colonel Price to Executive of that Brigade and brought Colonel Harold B. Parsons to command of the Fifth Marines. The writer tenders heartfelt thanks to these officers: the interest and encouragement they gave during the entire period of the writer's service as instructor to the Fifth Marines have proven most helpful in the writer's completion of his undertaking.

During the period of his service as instructor, the writer enjoyed the able assistance of Lieutenant James M. Masters, U.S.M.C., and Lieutenant William A. Kengla, U.S.M.C. These two latter named gentlemen were formerly pupils of the writer in Individual Combat at the United States Marine Corps Basic

School for Officers: they are both fine swordsmen. Being of inventive genius, Lieutenants Masters and Kengla devised several excellent new forms of attack and defense, as shown in this treatise.

Now come the very latest developments in the art of Defendu (originated by the celebrated Mr. W. E. Fairbairn, Assistant Commissioner, Shanghai Municipal Police), and of Jiu-Jitsu as shown by First Lieutenant Samuel G. Taxis, U.S.M.C., recently returned from Shanghai: there, in addition to his other Military duties, he was instructor in these arts. Following a series of conferences with Lieutenant Taxis several of his particularly noteworthy assaults are described in Part III of this Manual.

As a Professor on the Faculty of the Bureau of Investigation, the U. S. Department of Justice, the writer instructs in Individual Combat under Command of the World's greatest Conqueror of crime, The Honorable J. Edgar Hoover.

And the writer owes an especial debt of gratitude to the world-famous all-around athlete, Colonel C. J. Miller, now commanding the Fifth Regiment of Marines. Under Colonel Miller's personal instruction, the War Regiments were trained and developed in athletic prowess. The Colonel personally engaged in contests of strength and skill against the "pick of the men." As an American Bayonet Fighting Champion and as one of the boxing instructors of the Marine Corps, he discovered and trained several champions. The greatest of those whose training he encouraged

proved to be the only perpetually undefeated retired
heavyweight world's champion in the history of the
Ring, the mighty Marine, Gene Tunney, until recently
Captain in the Marine Corps Reserve. The friendship
of Colonel Miller has been an ever brilliant source of
inspiration to the writer.

In the summer of 1936 the writer was signally hon-
ored by again serving as Combat Instructor under the
distinguished command of Colonel C. J. Miller, who
had succeeded to the command of the Fifth Regiment,
United States Marine Corps.

During his terms of service as Combat Instructor
on the staff of the faculty of the Marine Corps Basic
School for Student Officers the writer has had the
privilege and advantage of serving, as Combat Instructor, under the successive brilliant commands of Colonel
Philip M. Torrey, Colonel A. D. Rorex, Colonel William
Dulty Smith, Colonel Julian C. Smith and Colonel
A. H. Turnage. Owing to the gracious personal interest
taken by each of these particularly able Officers, the
writer was at all times accorded every advantage
enabling him to develop and improve his work. He
received infinite inspiration and continual encouragement from the Commanding Officer at Headquarters
of the Marine Corps Schools, the distinguished Major
General J. C. Breckinridge, since and now in command
of the U. S. Marine Corps Department of the Pacific.

The writer has served in the United States Marine
Corps as Combat Instructor during consecutive terms
of office of the following Major General Commandants:

Major General George Barnett, Major General John
A. Lejeune, Major General Fuller and Major General
John H. Russell; and now he has the continued honor
to serve under the present Commandant of the United
States Marine Corps, Major General Thomas Holcomb,
the brilliant officer who formerly succeeded Major
General Breckenridge as the Commanding Officer of
the Schools of the United States Marine Corps.

PART I

BAYONET FENCING

BAYONET FENCING is a refinement in the use of the Bayonet, more scientific and effective than Bayonet Fighting. The Bayonet Fencer does not look upon his piece as a combination pike and mace, but as a "blade" of which the bayonet is the point. For this reason the bayonet fencer carefully guards his rifle against possible injury; he rarely uses the butt, relying habitually on his skill with the point. There are only four butt strokes that should ever be used: one is from the "Square Guard" position and another is the *up* stroke at the groin, directly following "Left Parry" as in the following command, "Left Parry, Butt strike, cut down, Pass by." The "cross-counter" "kick" of the rifle heel at the jaw is made by a straight arm blow; so is the butt stroke at the chest *directly* delivered with the heel of the piece. None of these four butt strokes imperil the rifle's good condition. The rifle *head guard against clubbed rifle is eschewed.* Such a guard tends to reduce one's rifle to kindling wood as it is the assault of the clubbed rifle which is swung from the barrel, the stock thus becoming the striking weapon. The bayonet fencer should meet such an attack by delivering his

(13)

point at the opponent's throat. Thus it will be seen
that the bayonet fencer is more definitely instructed

Old bayonet fighting "on guard" position, showing bayonet
edge downward.

in marksmanship than the bayonet fighter. The
bayonet fencer is instructed to keep his rifle clean
and in perfect condition for shooting at all times. He

should come through a bayonet charge with blood on
the blade but with the rifle unsullied and unharmed.

New bayonet fencing "on guard" position, showing bayonet
flat side up with cutting edge due right.

He should parry with his bayonet, and not with his
rifle, and deliver his *point* into his opponent as the
counter against a *swinging* or clubbed rifle attack.

THE "ON GUARD" POSITION OF THE BAYONET FENCER

While the stance of the bayonet fencer in the "on guard" position is similar to that prescribed in the ordinary bayonet course, there is one distinct difference. The bayonet fighting position is rigid, but absolute elasticity must be had in the fencer's "on guard" position. Pictures on pages 14 and 15 alternately show the incorrect and the correct positions herein prescribed. The bayonet point must be presented to the opponent with the blade flat and the edge directly to the right (as blades of every type are scientifically presented towards an opponent). Pursuant to the fencing blade position, the butt of the rifle rests laterally against the holder's crooked under-elbow and forearm. A blade attack from this lateral position is much more difficult and almost impossible to parry: it is the more powerful thrust. Furthermore, if the blade enters flatly between the ribs it can be readily withdrawn, whereas, if it is driven into the body perpendicularly it is apt to become caught or wedged between the ribs and be difficult to withdraw. Close attention is urged to the student's studying the pictures, pages 14 and 15, to learn the necessary ease and grace of the bayonet *fencer's* position. If, perchance, the extended left hand or arm is wounded and it is incapacitated, the rifle's position is still maintained by its secure hold of the supporting right forearm and grasped right hand. The left foot is advanced about sixteen inches in front of the right foot. As in sword fencing or boxing the feet must not be too far apart to impede rapid movements in fencing, shifting front, or rear pacing, or wide stepping.

The "Square Guard" Position
Command: "Square Guard"

This is best taken from "*On Guard*" position; forward
foot steps back, on line with stationary rear foot, to
a straddle stance, and rifle is simultaneously carried,
with "On Guard" grips maintained, to horizontal
position four inches below chin, barrel down. This
should clear space in a crowd.

Point and Butt from Square Guard

Commands: "Point and Butt"; "Butt and Point";
"To the left, point and butt"; "To the right, butt and
point"; "To the rear, butt and point," or "To the rear,
point and butt." In the latter two commands, the
turning direction is designated by the first named
assault, butt or point. A short step-in or a short
jump-in should be executed with each of the foregoing
commands.

Juggling the Piece

The commands: "Guard." "Short Guard." "Jab
Guard." "Guard."

From the "On Guard" position, the rifle is quickly
thrown by both hands simultaneously into the grasp
of "Short Guard" position. The rifle is again thrown
into the grasp of "Jab Guard" position when the right
foot is brought up directly to the rear of the advanced
left foot, and a slight crouch is taken. From this
position the rifle is quickly thrown back into the grasp
of "On Guard," and the crouch is changed to the

"On Guard" position with the right foot about sixteen inches rear.

Command: "Pass, shift—Parry and Point." The foregoing described shifts in Guard are each in turn executed under a repetition of the latter command, each part of the command being executed as each particular part of the command is given. The *"pass"* here ordered is a *"front-pass"* described in "Steps" (page 30), and is repeated with each shift in "guard."

"LEFT GUARD"

At this command, from the customary "On Guard" position, the left foot steps sixteen inches behind the right foot as the rifle is quickly thrown to the *left* side grasp of right hand at balance and left hand on small of stock. Thus the *left* guard position, although opposite, resembles the stance of the customary *"on guard"* position to the right, and the left forward hand is relieved from being further hand-cut.

Command: "Left Parry—Butt Strike—Cut Down— Pass By." This movement is especially prescribed for an advancing wave in a bayonet attack at close quarters.

Each particular movement is more violently made with a step-in or a leaping shift of the feet. Example: The left parry is executed from the "On Guard" position, with the body stationary. The Butt Strike, immediately following, is made directly at the groin and is a short, direct uppercut of the butt. This is better executed by a right *step-in* or a leaping shift of the feet.

"Cut Down"! directly follows with a left foot step-in,
or—better still—another leaping foot shift. In practice, the three sets of movements can be consecutively
taken: namely, the first set of movements with the
step-in with each consecutive movement; the third set
of movements, which are the best, are each taken with
a leaping shift of the feet.

BAYONET "GAIN AND POINT"

This is the new movement in bayonet fencing adapted
by the author from the "Gain and Point" of the épée.*
It will be found a highly effective bayonet movement.
The initial movement is taken from the stance of
"On Guard" by an exaggerated violent point at the
opponent's lower front middle section. The subsequent
success of this preliminary move will be principally due
to bringing, at the moment of the feint, the right foot forward directly back of the left, unnoticed by the adversary. This can be done by riveting the adversary's
attention on the "low point" by the violence of this feint
thrust. The right foot was concealed by the still stationary left or forward foot, and the opponent who aims to
parry the low thrust will scarcely realize that he
is menaced by an impending *throat* thrust (picture,
p. 20). The latter is speedily accomplished by avoiding blade contact from the attempted parry and

* "This sword movement, taught by Major William J. Herrmann, P.M.T.C., to Mrs. Dewar was repeatedly applied by
that lady in her match in New York against the Women's World
Foil Champion. Mrs. Dewar defeated the Champion by repeated
application of the "Gain and Point."

"Gain and point low." Showing *advancing* rear foot *gain* with feint low, to prepare for throat thrust.

"*Gain* and point low, *thrust high*" (2d move). Throat thrust from pace gained on opponent who failed to contact his low parry.

making an instantaneous forward lunge step of the left
foot, accompanied by a gliding thrust at the throat as
shown in picture, page 21. The success of this movement
actually depends upon the proper final execution of the
gliding blade, because it is required that no final jerky
indication of the throat thrust shall be given as this
would immediately bring the opponent's blade up to
the high parry. The entire execution of the final
move, after the attacker's feint thrust has drawn but
avoided the parry, if instantaneous, will find a clear
road to the throat.

Illustrations, pages 14 and 15, picture the difference
between the old bayonet fighting position (page 14) and
the new bayonet fencing position (page 15). It will be
noticed that all blades with a cutting edge as recommended in the herein-shown guard positions of the
knife and bayonet are held with the flat side up and the
cutting edge directly to the right. This guard position
of the bayonet directly follows the stance of the French
guard position of the Broadsword, excepting that in
the latter "on guard" stance the right foot is advanced
and the left foot is rear, while the sword is correctly
held in the right hand with edge to the right. This
position of the blade insures free withdrawal of the
blade if it has been deeply thrust through the ribs
and into the opponent's body. The writer stresses these
instructions by returning to the subject of the bayonet
fencer's "on guard" position (pages 14 and 15). In
any event, the throat is recommended as the ultimate
target, although *feints* are more effectively executed
to the body. Danger of entangling one's bayonet in

the clothing of an adversary renders the thrust into the throat advisable, particularly because the throat is uncovered and the thrust there instantly fatal. The first two inches of the blade thrust is sufficient. *Through* thrusts, even at the body or any part of the anatomy, should be forbidden by the instructor. There should never be more than three inches of the blade into the body, or two inches into the throat, to insure instant withdrawal.

In and Out

This should be the slogan of every bayonet fencer, and the bayonet should be thrust and withdrawn with rapid successive movements in order that the bayonet fencer may be instantly prepared for attack or defense against other adversaries.

The Hand-cut

The chief movements prescribed for the bayonet and for the knife fencer are patterned from the sword, and are identical: It is for this reason that the bayonet "hand-cut" (illustration, page 24) and the knife "hand-cut" (illustration, page 25) are pictured next. From the edge of the flat "Guard" blade position the attacker right parries his opponent's blade and then instantly turns the edge of his own blade downward as he takes a left step to his opponent's right front side. As he steps he cuts the forward hand of his opponent. This movement is known as "The hand-cut," and the three movements necessary to its execution instantly follow

Bayonet right parry, left step and hand-cut.

(24)

Knife right parry, left step and hand-cut.

each other in succession, as prescribed in the command
"Right Parry, left step, and hand-cut."

AT THE THROAT—Identical Command

This defense, as recommended with reservations,
is primarily prescribed as the best guard against
the bayonet in the *trench*. The rifle is held in the
"Jab Guard" position and the blade and stock of the
piece, held point up perpendicularly, furnish a full
length guard to confront an enemy's bayonet attack.
In any event, this stance is recommended for the trench
"On Guard" position. Firstly, because the narrow con-
fines of the trench preclude free use of the piece in the
customary "On Guard" position. Secondly, because
the "Jab Guard" stance is most effective, at intimately
close combat, from which to deliver a telling thrust
upwards under the chin. The "Jab Guard" position
is the safest against a bayonet attack at the throat.
Present the flat blade, and parry with the edges: a
more powerful parry is thus ensured. It is recommended that this guard be frequently practiced against
a blunt or scabbarded bayonet. This is a comparatively easy and safe defense, even against a *series* of
thrusts at the throat. It should also be borne in mind
that the offensive bayonet is almost always held in the
old time bayonet fighter's position. This renders the
blade particularly easy to parry when it is thrust as
a top and bottom edged blade, as shown in the picture
of the old time guard position (illustration, page 14): it
is much easier to parry than the flat blade as presented
with sharp edge to right and recommended in this new
bayonet course (illustration, page 27).

"At the throat," attack from the new "Bayonet" position, which is harder to parry than the customary encountering flat blade with edge down. The *Jab guard* is here purposely pictured in the customary position; it cannot withstand the flat blade attack as shown. Therefore, present the flat blade defense and parry with the edges as prescribed in the accompanying text entitled "At the throat."

(27)

Throw Point—Identical Command

In bayonet combat or duel this old fencing movement cannot be improved upon to reach an opponent too far distant for a thrust from the guard position. As shown in the accompanying illustration, the right foot is advanced in front of the left and the point is thrown, with the blade flat above and below, sharp edge to right, into the adversary. At the same time the left arm is extended with the left hand free in the air beneath the middle of the stock, so that, when the throw is accomplished, the extended stock of the piece may be easily caught and the rifle restored to the necessary balance of the "on guard" position. (Illustration page 29.)

The Parry

This is executed with a powerful blade rap of the opponent's blade to right or left, or above or below.

Croisé

An excellent method of defense and attack prescribes that a right or left parry becomes a downward parry: this by an adroit wrist turn down of one's rifle-holding front hand. This "turn down" must not be telegraphed, but applied only at contact: it should imprison the opponent's blade, then cut his hand and make way for one's "reposte" into the throat.

The Knock Down

To knock down an opponent, parry right and instantly step in with the right foot, bringing the stock of your piece against your opponent's: then press forward against his stock and carry your left foot in the air outside and behind his forward left leg and kick violently, heel first, into the back of the calf of his left

The time-honored and unexcelled "throw-point."

leg: thus make him lose his footing and fall backward. The butt stroke at the chest, as described in BAYONET FENCING (page 13), should also be carried through to a "knock down."

THE DEFENSE.

Colonel C. J. Miller devises the following defense: the prostrate one can avoid the death thrust from his standing adversary if he successfully encompasses with his left instep the attacker's ankle behind the heel of the latter's forward foot, and sets his own right foot firmly against the upper front shin bone of the attacker directly below the latter's knee.

STEPS

Advance.—This commands a single left step and right step forward, retaining the "*on guard*" position.

Retire.—Opposite of the "*advance*" movement, prescribing a left step and right step *rear*, retaining the forward "*On Guard*" position.

Left Step.—This is most effectively made with an accompanying preliminary right parry, but in any case this is a left foot step *left*, instantly followed by bringing the right foot back of the left to "on guard" position.

Right Step.—A step to the right with the right foot followed by a coördinated step to the right with the left foot to "on guard" position.

Front Pass.—This commands a forward step of the right foot twelve inches to the front of advanced left foot, immediately followed by the advance of the left foot beyond the right foot to the "on guard" position.

Rear Pass.—This is a directly opposite movement to the front pass, viz., the passing of the left foot twelve inches to the rear of the right foot immediately followed by the passing of the right foot to the rear of the left foot so that the proper "guard" position is resumed.

Leaps

A leap is taken directly from the "guard" position
with the rifle bearing bayonet *thrust* violently forward
into the opponent's middle section. The spring-off of
this leap can be well taken from the rear leg when it
takes the second step in an *"advance."* The FRONT
PASS AND LEAP is by far the best and most effective,
and most terrifying to the adversary, and the leap
should be taken directly following and from the forward
step of the right foot.

Volt

Each "Volt" command is preceded by the words,
"Right," "Left," "To the rear, right" or "To the
rear, left." The volt is executed on the ball of the
forward foot, carrying the rear foot around to conform
with the "on guard" position. During every *volt* the
rifle barrel must be raised perpendicularly (in order to
clear intervening objects) and lowered for point attack
instantly on arrival in the new "on guard" position.

PART II

KNIFE FIGHTING

CONSIDERABLE space in this treatise is given to knife fighting, because the Marines serve in many knife fighting countries and are frequently called upon to capture or fight against the dagger, machete or bolo. There are countries in Asia, Europe, Central America, Africa, and South America where the knife is a chief fighting weapon. While the military police in such countries, if they be Marines as is sometimes the case, can hardly attempt to match skill in the use of the bolo, machete, dagger or other type knives of the native, they can *draw* the bayonet and apply the hand-cut which is an unknown art to the *native* knife men. The hand-cut is particularly prescribed for use with the *bayonet as knife* and is an exquisitely scientific movement, taken from the sword and known to few others than scienced swordsmen. The skilled épée fencer or duellist thrusts at the sword hand and arm of an opponent; the scienced broad-swordsman *cuts* or thrusts at the sword hand and arm. When time does not permit the attachment of the bayonet to the rifle, or when the bayonet is worn in the belt and no rifle is carried, it is prescribed to use the bayonet as a disarming weapon against the armed adversary. In fact, with a quick cut to the opponent's knife-holding hand, it

(33)

is possible for the bayonet thus used to disarm several in a group of attacking knife men. There are various methods of wielding the knife in the many respective countries where the dagger is publicly and generally recognized as a standard weapon, and the overhand guard and stroke and the underhand guard and stroke are separately characteristic to particular races and are standardized and correct. Notwithstanding, the infinitely *superior* stance and method of the truly scientific knife duellist traces directly back to Roman Amphitheatre days; then the dagger duellist fought to the death. The best of these knife fighters are recorded to have been Gauls, who had been made slaves, as the gladiators were in ancient Rome. These old-time gladiators used what is still today the accepted method of the large majority of professional or champion knife duellists. The names of the movements are Gaelic-Roman. Underhand or overhand dagger contestants confronting the cool skill of the prescribed dagger *duellist* would be at a disadvantage like the *amateur* boxer facing the professional. Hand cutting is a practically unknown art to the underhand or overhand dagger fighter, and the straight knife-hold stance of the skilled duellist places the underhand or overhand dagger fighter at a disadvantage.

But, while the Gaelic-Roman names for the knife movements are still used, the following course of instruction teaches the use of the knife as prescribed by the late Colonel James Bowie, U. S. A. The Bowie knife has proved the most complete knife and knife method. While the Colonel traced his methods of attack and

defense through the lines of knife history as recited in
this brief preamble, the following course of instruction
is after the teachings of the Bowie knife as prescribed
by the Colonel himself: he was a celebrated sword duellist. The knife had its inception when Colonel Bowie
broke his sword in a duel and continued his fight by
closing in and killing his opponent with the shortened
broken blade which he still held at the hilt. Thus his
newly found weapon was fashioned as a straight blade
of the precise length of the broken blade with which
he killed his enemy. Not only did he prove with his
newly found blade to be the greatest knife fighter of
his time, but it is related that when he was ill in bed
he was attacked by some nine Mexican Indians, who
stole in upon him to take his life with tomahawks and
knives. From his sick bed Colonel Bowie met their
united attack with his Bowie knife: with this he killed
seven of the Indians before he himself succumbed. It
is related that he was found dead in bed with the bodies
of seven dead Indians about him. The other two
Indians of the attacking party fled after receiving
wounds from the Bowie knife. This is one story of
his death, but another account tells that he was killed
with Colonels Travis and Prockett during the taking
of the Alamo. Colonel Bowie was born in Georgia in
1790 and met his death in Alamo, Texas, March 6,
1836. Although he settled in 1802 in Chatahoula
parish, Louisiana, with his brother and parents, he
later emigrated to Texas: there he took a foremost part
in the Texan Revolution. He opposed the Mexicans
in battles during the year 1835, and eventually com-

manded his troops as Colonel. The name of Colonel
Bowie as a soldier and a fighter is immortalized.

As is elsewhere recounted in this Manual, many grad-
uates from the U. S. Marine Corps Student Officers
Basic School continue their study and practice in individual combat. They frequently return to the School
and tell of subsequent experiences. An outstanding
example was related at the Basic School by a prominent
Marine Aviator: he said that he and a fellow officer
had continued their individual fighting practices and
that each always carries a bayonet in his belt.

In Nicaragua the two drew their bayonets against
an attack of the enemy and successfully hand-cut their
way to safety through this force of some twenty
Machete fighters. He testified that the knowledge of
knife *Science* saved their lives. Thus, two Marine
skilled knife fighters defeated twenty Machete fighting
opponents.

In Germany the Army officers, the police and the
Hitler Storm Troopers are now all armed with the
knife which they use as either knife or bayonet.

On Guard

Command: "On Guard"!

The correct guard position of the dagger is shown in
illustration, page 37. In taking this photograph, the
camera was held on the ground to fully show the blade.
Unfortunately, the picture taken from this position
looks as though the blade were pointing slightly upward. As a matter of fact, this was not and is not the

case; the blade must be positively held strictly level on the opponent's middle so that no opening is presented for a hand-cut or thrust from said opponent's

Correct knife "on guard" position, showing ever ready left fencing arm balance and grab hand.

knife. It will be seen that the grab hand is ever ready to apply as shown in picture, page 39. This picture represents what is known as the—

Outside Parry and Grab—Identical Command

To execute this movement the opponent's blade is parried from the outside, and instantly afterwards the wrist of his knife hand is grasped from the outside by the disarming (left) "grab hand" of the defensive opponent who parried (illustration, page 39).

Inside Parry and Grab—Identical Command

This movement is not pictured, but is the opposite of the "outside parry and grab" as shown in the accompanying illustration. The wrist of the opponent's knife-holding hand is grasped from the inside in similar fashion immediately following a parry of his blade from the inside—it is the opposite side "parry and grab" of the picture here shown.

The Parry and Grab Follow-up

Wrist grabs are taken with fingers up, thumb down. And immediately following his left hand grab, the defense steps in with left foot advance.

Blade Position

As prescribed in the bayonet stance, the knife is *also* held with the flat side above and below, and the cutting edge facing outward to the right. The knife hold is correct when, palm down, the forefinger of the knife-holding hand encircles the bayonet button at the handle. Whether held with the left hand or the right hand the blade should be held outward so that in either case

"Outside parry and grab." Showing an accomplishment of the ever ready left *grab* hand.

(39)

the forefinger of the blade-holding hand presses against
the button at the bayonet handle. The position of
the blade as a detached knife or as a bayonet on the
rifle is identical with the position of the blade of the
French broadsword *guard* position. As the hand- or
wrist-cut or thrust is the basic plan of attack in both
bayonet and dagger, so it is the *basic* attack of the
épée swordsman, and it is also a particularly effective
attack of the broadswordsman. In point of fact, it is
the particularly scientific attack known to best *swordsmen* and rarely known to bayonet fighters or knife
men. The ordinary bayonet fighting course does not
teach the hand-cut, and the usual stab and slash
dagger man knows nothing of this scientific play. The
natural skill and celerity of the bolo or machete in
native hands is definitely offset by the hand-cut which
is a swordsman's science.

EXTEND LEFT ARM REAR IN RIGHT THRUSTS

Always follow the swordsman's method of throwing
out your left arm straight rear when making a right
hand thrust; it adds velocity and balance. See illustration *Stoccata*, page 43.

IN-QUARTATA

Command: "In-quartata—Time—Thrust"!
To accomplish the In-quartata thrust, step with the
left foot to the rear and right of the right foot as shown
in the foot position of illustration, page 41. But in the
precise in-quartata movement the left step right rear is
accompanied by a *quarte* thrust at the lower body

"In-quartata." Defense's left step right rear and hand-cut.

(41)

of the opponent which the changed thrust position has place unguarded, "out of line." Illustration, page 41, while actually showing the in-quartata step, pictures a cut at the opponent's right wrist, and this precise movement was devised by Lieutenant Kengla.

The opposite of the in-quartata movement is called stoccata and consists of a left step to left and thrust to lower right body as shown in illustration, page 43.

PASSATA SOTTO

Command: "Passata Sotto—Time—Thrust"!

This movement is executed on an opponent who lunges forward with a high thrust. It is so graphically illustrated in the accompanying picture that a detailed description seems unnecessary. Here the more skilled knife fighter avoids the thrust of an adversary by stooping to his own left under his adversary's out-stretched arm and bringing the dagger point to the middle section of his adversary (illustration, page 44).

UNARMED DEFENSE AGAINST OVERHAND DAGGER ASSAULT

The accompanying illustration shows how the unarmed man may successfully defend himself against the overhand dagger thrust. This particular defense is prescribed by Major William J. Herrmann, P.M.T.C., former World's Bayonet Fighting Champion. The faster and more violent the attack, the easier this defense is of accomplishment. A quick upward jolt with the left hand at the elbow of the attacking arm

Stoccata—Left step and body thrust.

(43)

"Passata Sotto."

(44)

completely deflects and throws aside the attacker
(illustration below).

The writer especially recommends the favorite

The Major Herrmann defense against overhand knife attack.

unarmed defense which Colonel C. J. Miller, U.S.M.C.,
Bayonet Fighting Champion, prescribes against the
overhand dagger assault. The following is Colonel
Miller's own celebrated instruction in this movement:

"Catch the blow of the opponent's descending right forearm on your left bent forearm, step in quickly and pass your right arm in rear of the opponent's right upper arm (knife arm), so that your right hand or fist rests in front of the opponent's right forearm just above the elbow—then bend the opponent backwards, breaking the arm."

The Chair Sword Contest

As shown in the accompanying picture, this is the particular exhibit of a contest in the master weapon, the sword. This illustration is presented to more clearly show why the sword is the master weapon of all the blades, and how a complete defense can be had by skillful wrist movements of the seated sword scientist (illustration, page 47): he must score his point with the reposte.

At the Cercle Hoche in Paris, where the author frequently fenced with men many years older than himself, he recalls the special skill of a Monsieur Priam, an elderly gentleman more than seventy-two years of age, who, at the time, continued to be one of the great foil professionals of France. In his fencing bouts he scarcely ever found it necessary to take a single step, for he could hold an adversary at bay with exquisite sword play from his scienced wrist, and he scored his point with a "reposte" from his parry of the opponent's lunge.

The chair *sword* contest.

PART III

JIU-JITSU* AND SAVATE

THE selection of the very few Jiu-Jitsu movements prescribed in this course is particularly made along the lines of least effort, and the movements are such that require little or no strength, but only quickness of thought and action. In fact, they are all Jiu-Jitsu movements such as a quick-thinking, able-bodied woman can readily be taught to use. For instance, when one is attacked by a double handed grasp on one's throat, the intended victim's own hands should be immediately clasped and brought violently up together between the extended arms of the throttler. This will instantly disengage the throttler's grasp and throw his arms out of line; then the defense cups his hands and simultaneously claps the ears of his assailant. Such a counter-attack will likely break the ear drums of the marauder. Another defense is to seize a finger of the throttler and break it.

All Jiu-Jitsu wrestling movements that require particular science in trying for complicated holds or grasps are eschewed. The Jiu-Jitsu movements herein shown are strongly advocated for use at close quarters either when weaponless and confronting an armed opponent

* Jiu-Jitsu—Japanese, *JuJutsu:* freely translated, skill or dexterity (Jutsu) employed without fighting instruments (Ju).

(49)

or when holding a weapon in one's own right hand.   The
left hand is the particular one for assault in Jiu-Jitsu.
The fact that Jiu-Jitsu assaults leave the right arm

"Eyes out."

free to wield a weapon is clearly shown in two particularly effective Jiu-Jitsu movements, "Break the Wind-
pipe" in illustration, page 51, and "Eyes Out" as illustrated above.   In both these movements the entire

forearm must be directly in line with the outstretched hand. The point of the drive should come entirely from the forward thrust from the biceps and shoulder.

"Break the wind-pipe."

The delivery of the blow in this way is required to make it successful. The fingers and wrist must be rigid. Delivering the throat attack, as prescribed in picture above, will sever the windpipe; and in the "Eyes Out" assault, illustrated in picture, page 50, the first and second fingers are passed into and through the eyes.

The defense against "Eyes Out" prescribes one's own wide open hand held perpendicularly, outer edge forward, thumb in and against the nose between the eyes.

DEFENDU AND JIU-JITSU

First Lieutenant Samuel G. Taxis, whose initial lessons in Jiu-Jitsu were taught him by the writer, now brings to the Marine Corps the Science of the celebrated Mr. W. E. Fairbairn, originator of "Defendu." Lieutenant Taxis has instructed a Battalion of the Fourth Regiment of Marines in Jiu-Jitsu and Defendu. This the Lieutenant did in China where he trained and managed the Marine Boxing Team that won the Boxing Championship of China.

Lieutenant Taxis also took part as an instructor with Mr. Fairbairn, in teaching Jiu-Jitsu and Defendu to 200 Sikh police. There were few defenses against Jiu-Jitsu attacks before Mr. Fairbairn entered the field, but Lieutenant Taxis now shows a perfect defense against every one of the innumerable Jiu-Jitsu "holds" and "blows."

In the "Eyes Out" attack, the movement of Lieutenant Taxis requires less accuracy than is needed to execute the attack in picture, page 50. The heel of either right or left hand is placed against the opponent's chin, and the fingers are pressed or scratched into the opponent's eyes. This movement can be most adroitly accomplished by applying it to the upper hand in the Colonel C. J. Miller attack (see picture, page 66).

While Lieutenant Taxis plans to write a book on developments of Defendu and Jiu-Jitsu as he personally presents them, the writer cannot refrain from describing several particularly effective blows as taught by Lieutenant Taxis.

Lieutenant Taxis delivers all his Defendu and Jiu-Jitsu blows from the outer edge of the wide open stiffened hand, which he uses like a weapon.

He prefers this attack at the throat instead of the straightened finger attack as shown in "Break the Windpipe" (illustration, page 51).

By striking a person with the outer edge of the hand a smart blow in front, directly below the ribs, the Solar Plexus is reached: a similar smart blow from above between the neck and shoulder can break the collar bone; the *vital* blow is delivered at the point over the thinnest bone where the nose joins the head between the eyes. The bone here is as thin as paper, and a blow downward directly breaking this bone causes a brain hemorrhage which brings blood poison in the brain and death within sixteen to twenty hours.

Lieutenant William A. Kengla, who was a former pupil of the author in individual combat, has become an expert in Jiu-Jitsu. He has shown genius in the application of Jiu-Jitsu movements; and, along the line of Jiu-Jitsu, he has developed several movements. Illustration, page 54, shows graphically the manner in which an unruly person can be effectively handled. The man who would take another man captive catches, with his own right hand with fingers up and thumb down, the other's *left* hand fingers from behind, and

brings the hand forward so that the now unruly's arm
is bent at right angles. He grasps the biceps above the
elbow with his left hand and brings the unruly's

Lieutenant Kengla takes captive.

bended elbow-joint directly into and against the recep-
tacle of his own bent right arm. Retaining his grasp
of the outstretched fingers, he can then completely
control the further movements of a captive or lead him

to a place of detention by bending the captive's wrist
inward with his capture finger hold. (See picture
below.)

Conducting the captive.

PISTOL DISARMING FROM THE REAR

Situation.—You are caught by an opponent behind
you, with the barrel of his pistol in your back. Your

hands are up at his command, "Hands up, or I'll shoot"! or "Move, and I'll shoot"!

Action.—Keep the elbows closely touching the sides

"Hands up"!

of the body and elevate the hands as in picture above. Under no circumstances let the elbows leave the body or elevate the hands higher than the picture shows. Following illustration, page 57, turn quickly to the left, hitting the opponent's wrist with the left elbow. This must not be in any manner a push, but must be an actual blow of the elbow. Make this blow a distinct

"Move and I'll shoot"! Defense throws pistol "out of line" by striking assailant's wrist with left elbow.

movement, instantly following it by a left arm revolution of the opponent's right arm into the position as shown in illustration, page 58. The revolution of the

arm will reverse the elbow-joint so that good pressure
will break the arm and, in the strain suffered by the
opponent, his pistol can be easily taken by the right
hand of the defense.

The arm revolution and break, and pistol disarming.

PISTOL DISARMING FROM THE FRONT

When the assailant presses the muzzle of his pistol
in front against his intended victim's middle and says,

"Hands up, or I'll shoot"! the intended victim is strictly cautioned to elevated his arms precisely as shown in picture below, and *no higher*, elbows pressed against sides. In spite of the enemy's further warning,

"Hands up or I'll shoot"!

"Move and I'll shoot"! the intended victim is then advised to whip his left hand down, fingers up and thumb down, to a tight grasp of the enemy's pistol-hand wrist, and sweep the hand along to his own right in order to deflect the shot of the attacker. Many tests of this move have proven it to be completely effective.

Left hand grasp and right sweep of gun-man's pistol-hand wrist, in response to threat, "Move and I'll shoot"!

The enemy will invariably pull the trigger, but is rarely successful in shooting the victim. Illustration above shows the intended victim about to deliver a right

hand punch to the jaw of the marauder. A better
follow-up movement, which is directly prescribed in
Jiu-Jitsu is, with left grasp still on wrist, to take an

The Lieutenant Kengla knee to the crotch "follow-up" in pistol
disarming.

instant grasp with one's right hand on the opponent's
pistol-holding hand. Take the grasp with the forefinger placed directly on top of the assailant's trigger

finger: By bending the pistol-holding hand inward at the wrist, and suddenly pressing the trigger finger of the enemy, he is made to shoot himself. Illustration, page 61, of this *front* pistol disarming series presents an excellent follow-up movement, devised by Lieutenant Kengla: the knee is brought instantly up, Jiu-Jitsu fashion, into the crotch of the adversary.

An excellent rebuttal to the Jiu-Jitsu disarmament of the pistol onslaught is furnished by Colonel Julian C. Smith, United States Marines. The Colonel suggests:

First: In holding up a man with a pistol, keep at least three paces distance from him. If this is not possible, and the pistol must be held within his reach, instead of attempting to pull the trigger at his first movement, draw the right (pistol) hand smartly to the rear, avoiding the sweep of the opponent's left hand, and step back quickly with the right foot; bring the right forearm to a horizontal position with the wrist against the right side, keep the pistol pointed at the assailant (continuing the backward movement if necessary) and pull the trigger. At this close range the so-called "hip shot," which is really a waist shot, should be effective. The left hand is free to guard against the right hand punch to the jaw. Even if the pistol hand is grasped, the backward movement will tend to keep the pistol pointed at the opponent so the shot cannot be evaded; it also defeats the knee into the crotch movement.

ARM BREAK

Of the almost innumerable effective movements in Jiu-Jitsu and Defendu, none other can surpass in

immediate effectiveness the movement hereinafter
described. Left hand palm downwards grasp with
inner left thumb knuckle, pressing, between opponent's
right hand outside second and third finger hand
knuckles, and with one's second finger grasped around
and pressing inside of opponent's thumb. With this
grasp, his forearm must be brought into a position
directly at right angles with his upper arm. Now by
applying pressure the opponent's wrist must be bent
inwardly at an outside angle to opponent's forearm.
With one's left hand second finger and thumb applied in
positions as prescribed, the opponent's right hand wrist
and arm at the elbow-joint will be broken unless he
is quick enough to prevent the break by falling to
his right instantly to the ground.

GENTLE GRASPS

Be certain to always take *gentle* grasps with your
left hand and distract the opponent's attention by
right hand gestures until you have brought the opponent's right hand into position for the final twist: then
put instant pressure into your execution.

HAND SHAKES

To subjugate a man with the hand shake and with
the deception necessary in Jiu-Jitsu, take an especially
tight grasp of his right hand and elevate it and slip
the left hand underneath his right arm to a grasp on
the top of his left shoulder. Straighten your left arm
so that it comes directly below your opponent's right

elbow, and be careful to hold his palm up. Thus, by exerting pressure downwards in your grasp of your opponent's right hand you can readily break his arm at his elbow.

Or, by retaining your tightened grasp you can subjugate your opponent by quickly stooping and passing your head either to the right or to the left under the hand clasp. In either position to which you then arrive by straightening up, with your still retained hand clasp you have your opponent in an imprisoned position through the twists of his arm which have been brought about through your own movements.

The Finger Subjugation

When an opponent has grasped you around the middle to bend you backwards with his hands clasped behind you, straighten your forefinger and second finger and press them on his upper or lower lip. Here the nerve-centres around the lower and upper gums of the teeth are super-sensitive, and direct finger pressure will cause the strongest man to drop his hold. But all such finger pressure must be executed with the fingers straightened parallel with one's hand and arm. Pressure from bent fingers from a bent wrist is not effective.

To Lead a Person Out of the Room

With your left hand grasp the first three fingers of another's right hand, thumb inside of his fingers and lift his hand with his palm upwards with pressure downwards in your grasp on his fingers. He will submit to your authority.

Defense Against a Hand Push

If a man places the palm of his hand against you to push you backwards, place either or both of your hands against his hand and hold it and bend forward. Thus you can break his wrist.

Arm Thrust and Belt Hold

This is one of the best movements for disarming or taking prisoner, and it is comparatively easy of accomplishment. As in the Arm Thrust or Belt Hold (or attack) as shown in the accompanying illustration, the assailant or unruly offender is grasped quickly with the left hand under the front of the belt or the upper front of the trousers, while the heel of the right hand at the same time is instantly pressed upwards against the opponent's chin. The opponent succumbs (illustration, page 66).

Savate

Following a life's study of various standardized systems of Individual Combat among the world's peoples, certain movements appear noteworthy. Frenchmen fight with their feet in *Savate:* the toe kick is "taboo," and blow delivery is from the sole of the foot. One such blow is herein prescribed: it is aimed at the front shin bone directly below the adversary's knee. It breaks the knee-joint. Such a blow is recommended

in a fist fight when the assault drives one backward:
then duck down, bending backward to the right and
deliver the sole of the forward left foot as a blow.

Arm thrust and belt hold.

If threatened or attacked when seated, this identical
foot-blow with right or left foot on the standing adversary is instantly followed by a right or left "hook"
punch to his open jaw: his mouth invariably opens as

the man careens forward with invective surprise or renewed attack. The open jaw is easily broken. Several other *Savate* attacks are prescribed with Jiu-Jitsu in this Manual.

KNEE IN THE CROTCH OR BREAK THE INSTEP

This is *Jiu-Jitsu* or *Defendu*. When the assailant imprisons one's hands and arms, lift the knee violently into his crotch (illustration, page 61, "Pistol Disarming from the Front") or, again, stamp on his instep: a heavy stamp will break the instep.

FRISKING

This is another effective "foreign" assault. Straighten and stiffen the fingers and scrape their tips rapidly back and forth across the eyes and nose bridge of your intended victim. (Also see Jiu-Jitsu attack, "Eyes Out," illustration, page 50.)

REMEMBER!

You are *never* defenseless. The assailant's eyes are an easy mark. At close range a handful of gravel or any handy article might be thrown at the eyes, or a hat whipped into them.

Lieutenant Taxis prescribes that a handkerchief worn in the upper left-hand coat pocket can be loaded with a few buckshot sewed, in small bulk, into one corner. Such handkerchief can be seized out at the top edge by the right hand, and the loaded corner can be deftly flicked into the eyes of an assailant.

PART IV

BOXING

The only death-dealing play devised in boxing was invented by the late Robert Fitzsimmons, who, until Gene Tunney, was perhaps the greatest genius in ring history. Tunney was at all times merciful. Although himself a middleweight, Fitzsimmons held three world championships, being middleweight, light-heavyweight, and heavyweight world's champion.

As the writer was privileged to be one of the sparring partners of Fitzsimmons, that mighty fighter took especial pains to carefully instruct the writer in the intricacies of the boxing movements of his own invention. These were supremely remarkable but, strangely enough, the knowledge of them has not generally been carried down to posterity.

He developed one punch which was sure to kill if landed with killing intent, but, with such dangerous knowledge, Fitzsimmons had an unusually kind and sympathetic nature which forbade undue cruelty. In an encounter he was long-suffering to a fault. In spite of this, few fighters withstood the Fitzsimmons punch, and he fought, in all, 328 battles, of which he lost but 5, 2 of these being to the same man, the great James J. Jeffries.

(69)

THE KILLING SHIFT

This was the movement with which Fitzsimmons
scored his knockouts. With the shift he won the heavyweight championship of the world over the late wonder-
ful James J. Corbett. To execute the shift, a right
hook is aimed at the opponent's chin: at the same time
the right foot steps forward, adding speed and force
to the blow. This right step must land toes forward
with the heel or back of the foot securely placed directly
against and in rear of the opponent's forward left foot.
Then, if the onslaught of the right hook fails to land,
the assailant's body continues to follow the course of
his hook that missed. Here he must let the momentum
of his own unlanded right hooked punch carry him
down to the left until his left crooked forearm is at
right angles on the outside directly below his own left
knee. Then he straightens up his right arm, which he
swings quickly as a pivot to speed a left hand punch
which travels from below the outside of his slightly
bended left knee to the underpoint of the opponent's
chin. As he hurls this punch, he puts the entire weight
of his body back of it by straightening both knees
which he had bent to add the weight of his body to the
blow. While chin punches from a standing position
may break the jaw, the punch from underneath, if
delivered correctly and with full force, will drive the
upper jaw bones into the base of the brain and thereby
cause brain concussion which can result in death to
the victim. But Fitzsimmons used a preliminary blow
to pave the way for his "knockout" just described: it
places the opponent "off guard" and precisely posed to

receive the "finishing" punch. To accomplish this, execute the shift into the final position preceding the left hand jaw punch. Then, instead of punching the jaw, straighten up with a drive of one's left fist into the solar plexus. This organ can be reached by a punch-push into the opponent's front middle section directly below the ribs above the stomach. By driving the blow in deep to the solar plexus, the opponent is momentarily paralyzed: he will sag at the knees, drop his hands and droop, chin forward, into the exact position to receive the jaw "knockout." And now, be sure and go all the way back to the stooping posture, as first described, for delivery of the final punch to the jaw: ample time will be had if the solar plexus was reached, for thus the opponent is rendered temporarily helpless.

The late Stanley Ketchell told the writer that he had carefully copied the Fitzsimmons shifts in his own onslaughts, which he himself so brilliantly executed, but even Ketchell never obtained the knowledge of the combination which is herein disclosed. Ketchell gained his masterful hitting power by shifting the foot with each punch, but he did not carry his shifts through as herein described because of his incomplete knowledge of the Fitzsimmons method.

The writer most carefully instructed that celebrated athlete, Captain Alan Shapley, U.S.M.C., of an All-America Football Team, in the art of the Fitzsimmons shift. Captain Shapley became more adept in this particular style of boxing than anyone the writer has ever seen, since Fitzsimmons. He has scored many

knockouts as a boxer, and uses the Fitzsimmons shift
to perfection.

The Fitzsimmons side-step is the best, but it also

Bob Fitzsimmons side-steps.

is unknown among the boxers today (see picture
above). It furnishes the easiest and surest avoidance
of the peerless *left jab*. On the instant of an opponent's
left lead, his adversary pivots, heel left, on the ball of

his forward left foot so that his toes face directly to
the right. At the same moment, he lifts his right foot
an inch from the ground and replaces it. The logic

Bob Fitzsimmons shows the foul pivot blow.

that we think with the feet is thus proven true, as no
other movement is required to remove oneself out of
harm's way from the left jab. Ducking or parrying
are both too slow. The Fitzsimmons side-step, once

thoroughly mastered, is the surest and safest method
of avoiding the most devastating of all boxing attacks—
the *left jab*. This left jab was the favorite attack of the
late James J. Corbett, as it has since been of almost
all the world's most *expert* boxers. It won the World's
Championship for the great Champion of Champions,
the Marine Reserve Officer, Gene Tunney, against
Jack Dempsey.

THE NECK GRAB, RABBIT PUNCH AND PIVOT BLOW

The third invention of Fitzsimmons is especially
recommended for individual combat. Feint a blow
with either hand at the opponent's face, and, instead
of hitting, open the hand and grasp him at the back of
the neck. Immediately pull him quickly forward so
that his head is down. In his staggering position, slap
him on the back of the neck alternately with the open
palm of each hand until he becomes groggy. Then,
as you let his head up, a smash with the fist will "do
the trick." The outlawed pivot blow is excellent in
rough and tumble fighting. Directly following a frontal
attack, the combatant pivots right in a complete
circle, on the ball of his rear right foot and, with a
bended right arm, strikes his victim's face with the
outside of his crooked upper right elbow (see picture,
page 73).

www.ingramcontent.com/pod-product-compliance
Lightning Source LLC
Chambersburg PA
CBHW022044210326
41458CB00071B/126